Autonomous Vehicles

How Self-Driving Cars Work and What They Mean for Society

Roman F. Preciado

Table of Contents

Chapter 2: The Current State and Future Trends of Autonomous Vehicles

- How are autonomous vehicles classified and evaluated according to different levels of automation and performance?
- What are the current capabilities and limitations of self-driving cars in various scenarios and environments?
- What are the emerging and expected trends and innovations in the autonomous vehicle sector and how will they shape the future of mobility?

Chapter 3: The Social and Ethical Implications of Autonomous Vehicles

- How do autonomous vehicles affect the safety, efficiency, and accessibility of transportation systems and services?
- How do autonomous vehicles impact the environment, energy consumption, and emissions?

- How do autonomous vehicles influence the behavior, preferences, and values of drivers, passengers, and pedestrians?
- How do autonomous vehicles raise ethical, legal, and moral dilemmas and questions, such as responsibility, liability, and trust?

Chapter 4: The Policy and Regulatory Framework for Autonomous Vehicles

- What are the main challenges and opportunities for policy-making and regulation of autonomous vehicles at local, national, and international levels?
- What are the existing and proposed laws and standards for autonomous vehicles and how do they vary across different regions and jurisdictions?
- What are the best practices and recommendations for designing and implementing effective and inclusive policies and regulations for autonomous vehicles?

Conclusion

- What are the main findings and rit2 from the previous chapters?
- LWhat are the key implications and recommendations for various stakeholders, such as consumers, manufacturers, policymakers, and researchers?
- What are the main limitations and gaps in the current knowledge and understanding of autonomous vehicles and what are the directions for future research and development?

Acknowledgment

Introduction

Imagine a world where you can travel anywhere, anytime, without worrying about traffic, parking, or driving. A world where you can relax, work, or socialize while your car takes you to your destination safely and efficiently. A world where transportation is accessible, affordable, and sustainable for everyone. This is the world of autonomous vehicles (AVs).

Autonomous vehicles (AVs) are cars that can sense their environment and operate without human involvement. They are also known as self-driving cars, driverless cars, or robotic cars. AVs differ from conventional cars in that they use a combination of sensors, cameras, radar, lidar, GPS, and artificial intelligence (AI) to perceive, navigate, and control the vehicle. AVs can potentially offer many benefits and challenges for individuals, businesses, and society, such as improving safety, efficiency, accessibility, and mobility, while also raising ethical, legal, and social issues.

What are autonomous vehicles and how do they differ from conventional cars?

An autonomous vehicle is a car that can drive itself without requiring any input or supervision from a human driver. AVs can perform all the driving tasks that a human driver can do, such as following traffic rules, detecting and avoiding obstacles, changing lanes, parking, and navigating complex situations. AVs differ from conventional cars in that they rely on a variety of technologies and systems to enable autonomous driving, rather than on human perception, cognition, and decision-making.

What are the main components and technologies involved in self-driving cars?

The main components and technologies involved in self-driving cars are:

Sensors: These are devices that collect data from the external and internal environment of the vehicle, such as cameras, radar, lidar, ultrasonic, infrared, and microphones. Sensors provide information about the road, traffic, weather, obstacles, and the vehicle's position, speed, and orientation.

Computers: These are the hardware and software that process, analyze, and store the data from the sensors, and generate commands for the vehicle's actuators. Computers include central processing units (CPUs), graphics processing units (GPUs), field-programmable gate arrays (FPGAs), neural network accelerators (NNAs), and memory devices. Computers run various algorithms and models based on artificial intelligence (AI), machine learning (ML), computer vision (CV), and deep learning (DL) to enable perception, planning, and control of the vehicle.

Actuators: These are the mechanical or electrical devices that execute the commands from the computers and control the vehicle's movement and functions, such as steering, braking, accelerating, signaling, and locking. Actuators are connected to the vehicle's electronic control units (ECUs), which are the embedded systems that regulate the vehicle's subsystems and components.

Communication: This is the exchange of data and information between the vehicle and other entities, such as other vehicles, infrastructure, pedestrians, and cloud services. Communication can be achieved through wireless technologies, such as cellular, Wi-Fi, Bluetooth, and dedicated short-range communications (DSRC). Communication can enhance the vehicle's situational awareness, coordination, and safety, as well as provide access to additional services and features.

What are the benefits and challenges of autonomous vehicles for individuals, businesses, and society?

Autonomous vehicles can potentially offer many benefits and challenges for individuals, businesses, and society, such as:

Safety: AVs can reduce human errors, which are the main cause of traffic accidents, and thus save lives, prevent injuries, and reduce costs associated with crashes.

Efficiency: AVs can optimize traffic flow, reduce congestion, and improve fuel economy, and thus save time, energy, and money for travelers and transporters.

Accessibility: AVs can provide mobility and independence for people who cannot drive or have limited access to transportation, such as the elderly, the disabled, the young, and the low-income.

Mobility: AVs can offer more convenient, comfortable, and personalized transportation options and services, such as ride-hailing, ride-sharing, and delivery, and thus enhance the quality of life and well-being of users and customers.

Challenges:

Ethical: AVs can pose moral dilemmas and questions, such as how to balance the safety and rights of different stakeholders, how to assign responsibility and liability in case of accidents, and how to ensure fairness and justice in the distribution and access of AVs.

Legal: AVs can require new laws and regulations to address the legal issues and implications of AVs, such as how to define and certify the levels of automation, how to protect the privacy and security of data and communication, and how to enforce the compliance and accountability of AVs and their operators.

Social: AVs can impact the social and cultural aspects and values of society, such as how to manage the transition and integration of AVs with existing transportation systems and modes, how to cope with the potential displacement and disruption of jobs and industries, and how to foster the trust and acceptance of AVs among the public and stakeholders.

The purpose of this book is to provide a comprehensive and balanced overview of AVs, their history, technology, applications, implications, and challenges. The book covers both the technical and non-technical aspects of AVs and draws on the latest research and evidence from various disciplines and perspectives. The book does not aim to predict the future of AVs, but rather to inform and inspire the readers to think critically and creatively about the opportunities and risks that AVs present for the future of mobility and society. Fasten your seatbelts as we take you to the first Section: The History and Evolution of Autonomous Vehicles.

SECTION 1

The History and Evolution of Autonomous Vehicles

In this section, we will explore the origin and development of autonomous vehicles, or self-driving cars, from the early experiments in the 20th century to the current state of the art. We will also examine the key milestones and achievements that have shaped the field of autonomous driving, as well as the main players and stakeholders that are involved in the industry. By the end of this chapter, you will have a better understanding of how autonomous vehicles work and what they mean for society.

How did the concept and development of self-driving cars emerge and progress over time?

The concept and development of self-driving cars can be traced back to the early 20th century when inventors and engineers experimented with radio-controlled and wire-guided vehicles. Some of the earliest examples include:

- The American Wonder (1925), a radio-controlled car demonstrated by Francis Houdina on the streets of New York City.

- The Phantom Auto (1926), a Chandler car equipped with a transmitting antenna and operated by a person in another car that followed it and sent out radio impulses.

- The Firebird II (1956), a General Motors concept car that was described as having a **"brain"** that allowed it to move into a lane with a metal rod and follow it along.

However, these vehicles were not truly autonomous, as they still required human intervention or external guidance. The first self-sufficient and truly autonomous cars appeared in the 1980s, with the advancement of artificial intelligence, computer vision, and sensor technologies. Some of the pioneering projects include:

- The Navlab (1984), a project by Carnegie Mellon University that developed a series of autonomous vehicles using cameras, laser rangefinders, and neural networks.

- The ALV (1984), a project by DARPA and SRI International that developed a self-driving van using lidar, radar, and computer vision.

- The Eureka Prometheus Project (1987), a project by Mercedes-Benz and Bundeswehr University Munich that developed a self-driving car using cameras, radar, and computer vision.

Since then, numerous major companies and research organizations have developed working autonomous vehicles, such as:

- Google (2009), launched its self-driving car project (later renamed Waymo) and tested its vehicles on public roads.

- Tesla (2015), unveiled its Autopilot system, a semi-autonomous driving feature that uses cameras, radar, and ultrasonic sensors.

- Baidu (2015), which drove a retrofitted BMW 3 Series autonomously in Beijing using its Apollo platform, an open-source software and hardware system for self-driving cars.

- Delphi (2015), which drove an autonomous car from coast to coast in the US, covering 15 states and 3,400 miles.

- Uber (2016), which tested autonomous taxis with human co-pilots in Pittsburgh, using modified Volvo XC90s equipped with cameras, lidar, radar, and GPS.

What are the key milestones and achievements in the field of autonomous vehicles?

The field of autonomous vehicles has achieved several milestones and achievements over the years, such as:

- The DARPA Grand Challenges (2004, 2005, 2007), which were competitions sponsored by the US Department of Defense to foster the development of autonomous vehicles for military purposes. The challenges involved driving in urban and off-road environments, avoiding obstacles, and following traffic rules. The winners of the challenges were teams from Stanford University, Carnegie Mellon University, and Virginia Tech.

- The VisLab Intercontinental Autonomous Challenge (2010), was a 13,000-kilometer journey from Italy to China by four autonomous vehicles using only cameras and GPS, without any human intervention. The challenge took three months and crossed nine countries.

- The Nevada Autonomous Vehicle Legislation (2011), was the first law in the US to authorize the operation of autonomous vehicles on public roads. The law required a special license, insurance, and safety standards for autonomous vehicles. Nevada also issued the first autonomous vehicle license plate to a Google car in 2012.

- The Berkeley DeepDrive (2014), was a research center at the University of California, Berkeley, that focused on applying deep learning and computer vision to autonomous driving. The center collaborated with industry partners such as Baidu, Ford, GM, Honda, and Toyota, and produced several publications and datasets on self-driving cars.

- The Waymo One (2018), which was the first commercial self-driving car service launched by Waymo, the subsidiary of Alphabet (Google's parent company). The service operated in Phoenix, Arizona, and offered rides to a select group of customers using fully autonomous vehicles without human safety drivers.

Who are the main players and stakeholders in the autonomous vehicle industry and what are their roles and interests?

The autonomous vehicle industry involves various players and stakeholders from different sectors and domains, such as:

- Automotive manufacturers and suppliers, such as BMW, Ford, GM, Honda, Mercedes-Benz, Tesla, Toyota, and Volvo, which design, produce, and sell autonomous vehicles and components, such as sensors, actuators, and software. Their roles and interests are to gain a competitive edge, increase market share, improve customer satisfaction, and reduce costs and emissions.

- Technology companies and startups, such as Apple, Baidu, Google, Uber, and Zoox, develop and provide autonomous driving platforms, services, and solutions, such as software, hardware, data, and cloud. Their roles and interests are to innovate, disrupt, and capture new markets, customers, and revenues.

- Research institutions and academia, such as Carnegie Mellon University, MIT, Stanford University, and UC Berkeley, which conduct and publish scientific research and experiments on autonomous driving, such as algorithms, models, and systems. Their roles and interests are to advance the state of the art, solve challenging problems, and train the next generation of experts and leaders.

- Government and regulators, such as the US Department of Transportation, the National Highway Traffic Safety Administration, and the European Commission, which set and enforce the policies, standards, and regulations for autonomous vehicles, such as safety, testing, certification, and liability. Their roles and interests are to protect the public interest, ensure compliance, and promote innovation and development.

- Consumers and society, such as drivers, passengers, pedestrians, and communities, which use and benefit from autonomous vehicles, such as convenience, mobility, and accessibility. Their roles and interests are to demand and expect safe, reliable, and affordable transportation options, and to influence and shape the future of mobility.

In this Section we explored the origin and development of autonomous vehicles, or self-driving cars, from the early experiments in the 20th century to the current state of the art. We also examined the key milestones and achievements that have shaped the field of autonomous driving, as well as the main players and stakeholders that are involved in the industry. These are some of the historical and current aspects of autonomous vehicles and how they work and what they mean for society. However, these aspects are not static or fixed, but rather evolving and changing, depending on various factors and influences, such as technological innovations, scientific discoveries, social demands, economic incentives, and political regulations. Therefore, it is important to understand and appreciate the history and evolution of autonomous vehicles, and to recognize and anticipate the potential opportunities and challenges that they may bring for the future of mobility and society.

SECTION 2

The Current State and Future Trends of Autonomous Vehicles

In this section, we will investigate the current capabilities and limitations of autonomous vehicles, and the emerging and expected trends and innovations in the autonomous vehicle sector. We will also explain the different levels and types of automation and performance of autonomous vehicles, and how they are classified and evaluated. We will also describe the main components and systems of autonomous vehicles, and how they enable or support the operation and function of autonomous vehicles. We will also examine the various scenarios and environments where autonomous vehicles can operate, and the challenges and opportunities that they present. We will also look at the different applications and domains of autonomous vehicles, and how they affect or benefit the users and the society.

How are autonomous vehicles classified and evaluated according to different levels of automation and performance?

Autonomous vehicles (AVs) are classified and evaluated according to different levels of automation and performance, based on the degree of human involvement and intervention required in the driving task. The most widely used and accepted framework for defining and measuring the levels of automation is the one developed by the Society of Automotive Engineers (SAE) International, which has been adopted by the US Department of Transportation and other regulatory bodies. The SAE framework defines six levels of automation, ranging from Level 0 (no automation) to Level 5 (full automation), as shown in the table below.

Table

Level	Name	Description	Examples
0	No Automation	The human driver performs all aspects of the driving task, even when assisted by warning or intervention systems.	Conventional cars with anti-lock braking system (ABS) or emergency braking system (EBS).
1	Driver Assistance	The vehicle has a single automated system for either steering or acceleration/deceleration, while	Cars with adaptive cruise control (ACC) or lane keeping assist (LKA).

		the human driver performs the rest of the driving task.	
2	Partial Automatio n	The vehicle has multiple automated systems that can control both steering and acceleratio n/decelerat ion, while the human driver monitors the driving environme	Cars with Tesla Autopilot or Cadillac Super Cruise.

		nt and is ready to take over at any time.	
3	Conditional Automation	The vehicle can perform all aspects of the driving task under certain conditions, while the human driver is still required to respond to requests to intervene or take over when the system	Cars with Audi Traffic Jam Pilot or Honda Traffic Jam Assist.

		reaches its limits.	
4	High Automation	The vehicle can perform all aspects of the driving task under certain conditions, without any human intervention or supervision. The vehicle can also handle most situations by itself but may request human	Cars with Waymo Driver or Cruise Origin.

		assistance in rare cases.	
5	Full Automation	The vehicle can perform all aspects of the driving task under all conditions, without any human intervention or supervision. The vehicle can handle any situation by itself and does not need any human	Cars with no steering wheel or pedals, such as Zoox or Nuro.

		assistance.	

The levels of automation are not only determined by the technical capabilities of the vehicle but also by the operational design domain (ODD), which defines the specific conditions and scenarios under which the vehicle can operate autonomously. The ODD may include factors such as road type, speed range, geographic area, weather, traffic, and time of day. For example, a Level 4 vehicle may be able to drive autonomously on highways, but not on urban streets, only within a predefined area, or only during daylight hours. Therefore, the levels of automation are not necessarily linear or discrete but rather depend on the context and the use case.

What are the current capabilities and limitations of self-driving cars in various scenarios and environments?

Self-driving cars have achieved remarkable progress and performance in various scenarios and environments, thanks to the advances in artificial intelligence, computer vision, sensor technology, and connectivity. However, they still face significant challenges and limitations that prevent them from achieving full autonomy and reliability. Some of the current capabilities and limitations of self-driving cars are:

Perception: Self-driving cars use a combination of sensors, such as cameras, lidar, radar, and ultrasonic, to perceive and interpret their surroundings, such as road markings, traffic signs, signals, vehicles, pedestrians, cyclists, and animals. They also use computer vision and deep learning algorithms to process and analyze the sensor data and to detect and classify objects and events. Self-driving cars have demonstrated the ability to perceive and react to complex and dynamic situations, such as intersections, roundabouts, lane changes, and overtaking. However, they still struggle with some perception challenges, such as poor visibility, occlusion, ambiguity, novelty, and adversarial attacks. For example, self-driving cars may have difficulty seeing in fog, rain, snow, or darkness, or may be confused by objects that are partially hidden, look similar, are unfamiliar, or are deliberately manipulated to deceive them.

Planning: Self-driving cars use artificial intelligence and machine learning algorithms to plan and execute their actions, such as steering, braking, accelerating, signaling, and parking. They also use maps, GPS, and communication systems to navigate and coordinate with other vehicles and infrastructures. Self-driving cars have demonstrated the ability to plan and execute safe and efficient maneuvers, such as following traffic rules, avoiding collisions, and optimizing routes. However, they still face some planning challenges, such as uncertainty, variability, and ethics. For example, self-driving cars may have difficulty predicting the behavior and intentions of other road users or may encounter situations that require trade-offs or moral judgments, such as choosing between hitting a pedestrian or another car.

Human Factors: Self-driving cars interact with human drivers, passengers, pedestrians, and other road users, either directly or indirectly. They also depend on human operators, supervisors, and regulators, either remotely or locally. Self-driving cars have demonstrated the ability to interact and communicate with humans, such as using sounds, lights, gestures, and displays, to indicate their intentions, actions, and status. However, they still face some human factors challenges, such as trust, acceptance, and responsibility. For example, self-driving cars may have difficulty establishing and maintaining trust and acceptance from humans or may raise issues of responsibility and liability in case of accidents or malfunctions.

What are the emerging and expected trends and innovations in the autonomous vehicle sector and how will they shape the future of mobility?

The autonomous vehicle sector is undergoing rapid and disruptive changes, driven by technological, economic, social, and environmental factors. Some of the emerging and expected trends and innovations in the autonomous vehicle sector are:

Market Shift: The autonomous vehicle market is shifting from passenger cars to commercial vehicles, such as trucks, buses, and delivery vehicles. This is due to the higher profitability, lower complexity, and greater demand for commercial applications of autonomous vehicles, especially in the wake of the COVID-19 pandemic, which has boosted the e-commerce and logistics sectors. Several companies, such as Waymo, Tesla, Aurora, and TuSimple, are developing and testing autonomous trucks, which can reduce costs, increase efficiency, and improve safety for freight transportation. Similarly, several companies, such as Cruise, Nuro, and Starship, are developing and deploying autonomous delivery vehicles, which can offer convenience, accessibility, and sustainability for goods delivery.

Technology Innovation: Autonomous vehicle technology is innovating and improving in various aspects, such as perception, planning, and human factors. Some of the technology innovations are:

- **Sensor Fusion:** Sensor fusion is the process of combining and integrating data from multiple sensors, such as cameras, lidar, radar, and ultrasonic, to create a more accurate and robust representation of the environment. Sensor fusion can enhance the perception and performance of autonomous vehicles, by overcoming the limitations and complementing the strengths of each sensor type. For example, lidar can provide high-resolution and precise distance measurements, but is expensive and sensitive to weather conditions, while cameras can provide rich color and texture information, but are affected by lighting and occlusion. By fusing the data from both sensors, autonomous vehicles can achieve better object detection and recognition, especially in challenging scenarios.

- Reinforcement Learning: Reinforcement learning is a type of machine learning that enables autonomous agents to learn from their actions and experiences, by receiving rewards or penalties based on the outcomes. Reinforcement learning can improve the planning and decision-making of autonomous vehicles, by enabling them to adapt and optimize their behavior in complex and dynamic situations, without relying on predefined rules or human supervision. For example, reinforcement learning can help autonomous vehicles learn how to negotiate with other vehicles at intersections, or how to handle rare or novel events, such as road closures or emergencies.

- **Human-Machine Interface:** Human-machine interface (HMI) is the system that enables communication and interaction between humans and machines, such as sounds, lights, gestures, and displays. HMI can improve the human factors and user experience of autonomous vehicles, by providing information, feedback, and guidance to human drivers, passengers, pedestrians, and other road users. For example, HMI can help autonomous vehicles to indicate their intentions, actions, and status, such as turning, braking, or yielding, or to request or respond to human input, such as taking over or intervening, or to convey their confidence or uncertainty, such as alerting or warning.

Mobility Transformation: The autonomous vehicle sector is transforming the mobility landscape and creating new opportunities and challenges for various stakeholders, such as consumers, businesses, and society. Some of the mobility transformations are:

- **Mobility as a Service:** Mobility as a service (MaaS) is the concept of providing integrated and personalized mobility solutions and services to users, based on their needs and preferences, such as trip planning, booking, payment, and access. MaaS can leverage the potential of autonomous vehicles, by offering more convenient, comfortable, and affordable transportation options and services, such as ride-hailing, ride-sharing, and delivery. MaaS can also enable new business models and revenue streams for mobility providers, such as subscription, pay-per-use, or advertising.

- **Smart City:** A smart city is a concept of using information and communication technologies to enhance the quality and efficiency of urban services and systems, such as transportation, energy, water, waste, and security. Smart cities can benefit from the integration and coordination of autonomous vehicles with other smart devices and infrastructures, such as sensors, cameras, traffic lights, parking lots, and charging stations. Smart cities can also enable new features and functions for autonomous vehicles, such as platooning, vehicle-to-everything (V2X) communication, and smart grid.

- **Sustainable Development:** Sustainable development is the concept of meeting the needs of the present without compromising the ability of future generations to meet their own needs, by balancing the economic, social, and environmental aspects of development. Sustainable development can be supported by the adoption and diffusion of autonomous vehicles, by reducing the negative impacts and enhancing the positive impacts of transportation on the environment, society, and economy. For example, autonomous vehicles can reduce greenhouse gas emissions, energy consumption, and air pollution, by improving fuel efficiency, traffic flow, and vehicle utilization. Autonomous vehicles can also improve social equity, inclusion, and well-being, by providing mobility and accessibility for all, especially for the underserved and disadvantaged groups.

In This section we investigated the current capabilities and limitations of autonomous vehicles, and the emerging and expected trends and innovations in the autonomous vehicle sector. We also explained the different levels and types of automation and performance of autonomous vehicles, and how they are classified and evaluated. We also described the main components and systems of autonomous vehicles, and how they enable or support the operation and function of autonomous vehicles. We also examined the various scenarios and environments where autonomous vehicles can operate, and the challenges and opportunities that they present. We also looked at the different applications and domains of autonomous vehicles, and how they affect or benefit the users and the society. These are some of the current and future aspects of autonomous vehicles and how they can transform mobility and society in various ways. However, these aspects are not inevitable or deterministic, but rather contingent and dynamic, depending on various factors and uncertainties, such as technology, market, policy, and society. Therefore, it is important to monitor and anticipate the potential opportunities and challenges that these aspects may bring and to prepare and adapt accordingly.

SECTION 3

The Social and Ethical
Implications of Autonomous
Vehicles

In this section, we will analyze the social and ethical
implications of autonomous vehicles, and how they
impact or improve the quality of life and well-being
of the drivers, passengers, pedestrians, and other
road users. We will also discuss the effects of

autonomous vehicles on the safety, efficiency, and accessibility of transportation systems and services, and the trade-offs and choices that they involve. We will also investigate the environmental, energy, and emission impacts and challenges of autonomous vehicles, and how they challenge or contribute to the sustainability and resilience of the planet and the society. We will also explore the behavior, preferences, and values of the human agents involved or affected by autonomous vehicles, and how they shape or respond to the design and development of autonomous vehicles. We will also examine the ethical, legal, and moral dilemmas and questions that autonomous vehicles raise, such as responsibility, liability, and trust, and how they affect or ensure the accountability and transparency of autonomous vehicles.

How do autonomous vehicles affect the safety, efficiency, and accessibility of transportation systems and services?

Autonomous vehicles (AVs) have the potential to improve the safety, efficiency, and accessibility of transportation systems and services, by reducing human errors, optimizing traffic flow, and providing mobility and independence for all. However, they also pose some risks and challenges that need to be addressed and mitigated, such as cyberattacks, system failures, and social exclusion. Some of the effects of AVs on the safety, efficiency, and accessibility of transportation are:

Safety: AVs can reduce the number and severity of traffic accidents, which are mostly caused by human factors, such as distraction, fatigue, impairment, and aggression. AVs can also avoid or minimize the harm caused by external factors, such as weather, road conditions, and wildlife. AVs can use sensors, cameras, radar, lidar, and artificial intelligence (AI) to perceive and react to their surroundings, and to follow traffic rules and signals. AVs can also communicate and coordinate with other vehicles and infrastructures, using vehicle-to-vehicle (V2V) and vehicle-to-infrastructure (V2I) technologies, to enhance their situational awareness and safety. However, AVs are not immune to cyberattacks, system failures, or human interference, which can compromise their safety and security. AVs may also encounter situations that are unpredictable, ambiguous, or conflicting, which can challenge their decision-making and performance. For example, AVs may have to deal with pedestrians who jaywalk, cyclists who swerve, or drivers who run red lights.

Efficiency: AVs can improve the efficiency and productivity of transportation systems and services, by reducing congestion, travel time, and fuel consumption. AVs can optimize their routes, speeds, and distances, using maps, GPS, and navigation systems, to avoid traffic jams and delays. AVs can also form platoons, which are groups of vehicles that travel closely together, using V2V and V2I technologies, to increase road capacity and reduce aerodynamic drag. AVs can also enable new modes and models of transportation, such as ride-hailing, ride-sharing, and delivery, which can increase the utilization and availability of vehicles, and reduce the need for parking and ownership. However, AVs may also have negative impacts on the efficiency and productivity of transportation systems and services, such as increasing the demand and distance of travel, creating new sources of congestion, and displacing or disrupting existing modes and models of transportation. For example, AVs may induce people to travel more frequently, farther, or longer, or to use AVs instead of public transportation, walking, or cycling, which can increase traffic volume and energy consumption.

Accessibility: AVs can enhance the accessibility and inclusivity of transportation systems and services, by providing mobility and independence for people who cannot drive or have limited access to transportation, such as the elderly, the disabled, the young, and the low-income. AVs can also offer more convenient, comfortable, and personalized transportation options and services, such as ride-hailing, ride-sharing, and delivery, which can cater to the diverse needs and preferences of users and customers. AVs can also improve the affordability and availability of transportation, by reducing the costs and barriers of ownership, maintenance, and insurance, and by increasing the supply and competition of transportation providers. However,

AVs may also create or exacerbate the accessibility and inclusivity gaps and inequalities in transportation systems and services, such as discriminating or excluding certain groups of people, creating digital or physical divides, and undermining social or cultural values. For example, AVs may favor or privilege those who can afford or access the technology, or those who live in urban or developed areas, while neglecting or marginalizing those who cannot or do not, or those who live in rural or underdeveloped areas.

How do autonomous vehicles impact the environment, energy consumption, and emissions?

Autonomous vehicles (AVs) have the potential to impact the environment, energy consumption, and emissions, both positively and negatively, depending on various factors and scenarios, such as the level of automation, the type of fuel, the mode of operation, and the behavioral and policy responses. Some of the impacts of AVs on the environment, energy consumption, and emissions are:

Environment: AVs can contribute to environmental protection and preservation, by reducing the land use, noise, and waste associated with transportation. AVs can reduce land use, by decreasing the need for parking and road infrastructure, and by increasing road capacity and density. AVs can reduce the noise, by operating more quietly and smoothly, and by avoiding unnecessary braking and accelerating. AVs can reduce waste, by extending the lifespan and durability of vehicles and components, and by facilitating the recycling and reuse of materials. However, AVs may also have adverse effects on the environment, by increasing resource use, biodiversity loss, and pollution associated with transportation. AVs may increase resource use, by requiring more materials and energy for the production and operation of vehicles and infrastructures, such as sensors, computers, and communication systems. AVs may increase biodiversity loss, by expanding the road network and encroaching on natural habitats and ecosystems. AVs may increase pollution, by generating more electronic and hazardous waste, and by emitting more electromagnetic and thermal radiation.

Energy Consumption: AVs can reduce energy consumption and dependence, by improving fuel efficiency and diversification of transportation. AVs can improve fuel efficiency, by optimizing driving behavior and performance, such as avoiding congestion, idling, and speeding, and by forming platoons, which can reduce aerodynamic drag and resistance. AVs can also diversify the fuel sources, by facilitating the adoption and integration of alternative and renewable energy, such as electricity, hydrogen, and biofuels. However, AVs may also increase energy consumption and dependence, by increasing the travel demand and distance, adding more weight and complexity to vehicles and infrastructures, and requiring more power and cooling for the operation and maintenance of vehicles and systems.

Emissions: AVs can reduce greenhouse gas emissions and air pollution, by reducing fuel consumption and combustion of transportation. AVs can reduce fuel consumption, by improving fuel efficiency, as discussed above. AVs can also reduce fuel combustion, by facilitating the transition and penetration of zero-emission and low-emission vehicles, such as electric, hybrid, and fuel-cell vehicles. However, AVs may also increase greenhouse gas emissions and air pollution, by increasing energy consumption, as discussed above, and by shifting the emissions from the tailpipe to the upstream sources, such as the power plants and refineries, which may not be as clean or efficient as the vehicles.

How do autonomous vehicles influence the behavior, preferences, and values of drivers, passengers, and pedestrians?

Autonomous vehicles (AVs) have the potential to influence the behavior, preferences, and values of drivers, passengers, and pedestrians, by changing the way they interact with and perceive transportation. AVs can affect the attitudes, emotions, and expectations of people, as well as the norms, cultures, and identities of society. Some of the influences of AVs on the behavior, preferences, and values of drivers, passengers, and pedestrians are:

Behavior: AVs can change the behavior and actions of people, by altering the incentives and costs of transportation. AVs can affect the mode choice, travel frequency, travel distance, and travel time of people, by offering more convenient, comfortable, and affordable transportation options and services, such as ride-hailing, ride-sharing, and delivery. AVs can also affect the activity and productivity of people, by enabling them to engage in other tasks and purposes during travel, such as work, entertainment, or rest. AVs can also affect the risk and trust of people, by exposing them to new sources and levels of uncertainty and vulnerability, such as cyberattacks, system failures, or ethical dilemmas.

Preferences: AVs can change the preferences and tastes of people, by providing them with more personalized and customized transportation solutions and experiences. AVs can affect the utility and satisfaction of people, by catering to their diverse and dynamic needs and wants, such as comfort, safety, privacy, or fun. AVs can also affect the loyalty and switching of people, by creating new markets and competition for transportation providers, such as subscription, pay-per-use, or advertising. AVs can also affect the ownership and sharing of people, by reducing the costs and barriers of ownership, maintenance, and insurance, and by increasing the utilization and availability of vehicles.

Values: AVs can change the values and beliefs of people, by challenging and transforming the meanings and roles of transportation. AVs can affect the identity and status of people, by affecting their sense of self and social recognition, such as their skills, autonomy, or prestige. AVs can also affect the culture and norms of people, by affecting their shared practices and expectations, such as their etiquette, courtesy, or responsibility. AVs can also affect the ethics and morals of people, by affecting their judgments and decisions, such as their fairness, justice, or altruism.

How do autonomous vehicles raise ethical, legal, and moral dilemmas and questions, such as responsibility, liability, and trust?

Autonomous vehicles (AVs) raise ethical, legal, and moral dilemmas and questions, such as responsibility, liability, and trust, by introducing new actors, scenarios, and outcomes that challenge the existing frameworks and principles of justice, rights, and duties. AVs can affect the allocation and attribution of responsibility, liability, and trust among various stakeholders, such as manufacturers, operators, users, and regulators, and create new conflicts and trade-offs among different values and interests, such as safety, efficiency, and fairness. Some of the ethical, legal, and moral dilemmas and questions raised by AVs are:

Responsibility: Responsibility is the obligation or expectation to perform or account for one's actions, especially when they have consequences for others. Responsibility can be moral, legal, or social, depending on the norms and rules that govern the behavior and the relationships of the agents involved. AVs can raise questions and issues about the distribution and assignment of responsibility among different agents, such as:

- **Moral Responsibility:** Moral responsibility is the responsibility to act in accordance with moral principles and values, and to be accountable for the moral outcomes of one's actions. AVs can raise questions and issues about the moral responsibility of different agents, such as:

  - Who is morally responsible for the actions and outcomes of AVs, especially when they cause harm or conflict with moral norms?

o How can moral responsibility be shared or delegated among different agents, such as manufacturers, operators, users, and regulators, and what are the criteria and mechanisms for doing so?

o How can moral responsibility be ensured or enforced among different agents, and what are the incentives and sanctions for doing so?

o How can moral responsibility be evaluated or measured among different agents, and what are the standards and methods for doing so?

- **Legal Responsibility:** Legal responsibility is the responsibility to act in accordance with legal rules and regulations, and to be liable for the legal consequences of one's actions. AVs can raise questions and issues about the legal responsibility of different agents, such as:

  - Who is legally responsible for the actions and outcomes of AVs, especially when they violate or conflict with legal norms?

  - How can legal responsibility be allocated or transferred among different agents, such as manufacturers, operators, users, and insurers, and what are the procedures and instruments for doing so?

o How can legal responsibility be established or proven among different agents, and what are the evidence and procedures for doing so?

o How can legal responsibility be compensated or punished among different agents, and what are the damages and penalties for doing so?

- **Social Responsibility:** Social responsibility is the responsibility to act in accordance with social expectations and values, and to contribute to the social welfare and development of others. AVs can raise questions and issues about the social responsibility of different agents, such as:

o Who is socially responsible for the actions and outcomes of AVs, especially when they affect or conflict with social interests?

- How can social responsibility be balanced or traded off among different agents, such as manufacturers, operators, users, and regulators, and what are the criteria and mechanisms for doing so?

- How can social responsibility be communicated or expressed among different agents, and what are the channels and modes for doing so?

- How can social responsibility be recognized or rewarded among different agents, and what are the benefits and incentives for doing so?

Liability: Liability is the legal obligation or duty to pay for or remedy the damages or losses caused by one's actions, especially when they are negligent or wrongful. Liability can be civil, criminal, or administrative, depending on the nature and severity of the damages or losses, and the type and level of the legal authority involved. AVs can raise questions and issues about the determination and allocation of liability among different agents, such as:

- **Civil Liability:** Civil liability is the liability to pay for or remedy the damages or losses caused by one's actions, especially when they breach a contract or a duty of care. Civil liability can be contractual or tortious, depending on the existence and terms of a contract or a duty of care between the parties involved. AVs can raise questions and issues about the civil liability of different agents, such as:

○ Who is civilly liable for the damages or losses caused by AVs, especially when they result from a defect or a malfunction of the vehicle or the system?

○ How can civil liability be allocated or transferred among different agents, such as manufacturers, operators, users, and insurers, and what are the contracts and policies for doing so?

○ How can civil liability be established or proven among different agents, and what are the evidence and procedures for doing so?

○ How can civil liability be compensated or settled among different agents, and what are the damages and remedies for doing so?

- **Criminal Liability:** Criminal liability is the liability to be punished or sanctioned for the damages or losses caused by one's actions, especially when they violate a criminal law or a public order. Criminal liability can be subjective or objective, depending on the intention and knowledge of the agent involved. AVs can raise questions and issues about the criminal liability of different agents, such as:

  - Who is criminally liable for the damages or losses caused by AVs, especially when they result from a deliberate or reckless action of the vehicle or the system?

- How can criminal liability be attributed or imputed among different agents, such as manufacturers, operators, users, and hackers, and what are the criteria and mechanisms for doing so?

- How can criminal liability be established or proven among different agents, and what are the evidence and procedures for doing so?

- How can criminal liability be punished or sanctioned among different agents, and what are the penalties and measures for doing so?

- **Administrative Liability:** Administrative liability is the liability to be fined or regulated for the damages or losses caused by one's actions, especially when they violate an administrative rule or a public interest. Administrative liability can be individual or collective, depending on the scope and impact of the action involved. AVs can raise questions and issues about the administrative liability of different agents, such as:

- o Who is administratively liable for the damages or losses caused by AVs, especially when they result from non-compliance or a breach of the vehicle or the system?

- o How can administrative liability be allocated or transferred among different agents, such as manufacturers, operators, users, and regulators, and what are the rules and regulations for doing so?

o How can administrative liability be established or proven among different agents, and what are the evidence and procedures for doing so?

o How can administrative liability be fined or regulated among different agents, and what are the fines and regulations for doing so?

Trust: Trust is the confidence or belief in the reliability, competence, and integrity of another agent, especially when there is uncertainty, risk, or vulnerability involved. Trust can be interpersonal, institutional, or technological, depending on the nature and type of the agent involved. AVs can raise questions and issues about the formation and maintenance of trust among different agents, such as:

- **Interpersonal Trust:** Interpersonal trust is the trust between human agents, such as drivers, passengers, pedestrians, and other road users. Interpersonal trust can be based on personal or social factors, such as reputation, experience, or norms. AVs can raise questions and issues about the interpersonal trust of different agents, such as:

  - How do AVs affect the interpersonal trust between human agents, especially when they change the roles and expectations of drivers, passengers, pedestrians, and other road users?

  - How can interpersonal trust be fostered or enhanced among human agents, and what are the factors and mechanisms for doing so?

- How can interpersonal trust be communicated or expressed among human agents, and what are the channels and modes for doing so?

- How can interpersonal trust be verified or measured among human agents, and what are the standards and methods for doing so?

- **Institutional Trust:** Institutional trust is the trust in the organizations or authorities that govern, regulate, or provide the services or systems related to AVs, such as manufacturers, operators, regulators, or insurers. Institutional trust can be based on legal or professional factors, such as rules, regulations, or standards. AVs can raise questions and issues about the institutional trust of different agents, such as:

- How do AVs affect the institutional trust in the organizations or authorities involved in AVs, especially when they introduce new actors, scenarios, and outcomes that challenge the existing frameworks and principles of justice, rights, and duties?

- How can institutional trust be established or maintained among the organizations or authorities involved in AVs, and what are the rules, regulations, or standards for doing so?

- How can institutional trust be communicated or expressed among the organizations or authorities involved in AVs, and what are the channels and modes for doing so?

- How can institutional trust be evaluated or monitored among the organizations or authorities involved in AVs, and what are the standards and methods for doing so?

- **Technological Trust:** Technological trust is the trust in the devices or systems that enable or support the operation or function of AVs, such as sensors, computers, communication systems, or navigation systems. Technological trust can be based on technical or functional factors, such as performance, reliability, or security. AVs can raise questions and issues about the technological trust of different agents, such as:

o How do AVs affect the technological trust in the devices or systems involved in AVs, especially when they involve complex and uncertain processes, such as perception, planning, and decision-making?

o AVs can affect the technological trust in the devices or systems involved in AVs, by exposing them to new challenges and risks that may compromise their performance, reliability, or security. For example, AVs may face situations that are unpredictable, ambiguous, or conflicting, which can test the limits and capabilities of their perception, planning, and decision-making processes. AVs may also encounter cyberattacks, system failures, or human interference, which can disrupt or damage their devices or systems.

○ AVs can also affect the technological trust in the devices or systems involved in AVs, by creating new opportunities and benefits that may enhance their performance, reliability, or security. For example, AVs may use sensors, cameras, radar, lidar, and AI to perceive and react to their surroundings, and to follow traffic rules and signals. AVs may also communicate and coordinate with other vehicles and infrastructures, using V2V and V2I technologies, to increase their situational awareness and safety.

How can technological trust be built or improved among the devices or systems involved in AVs?

- Technological trust can be built or improved among the devices or systems involved in AVs, by adopting and applying rigorous and transparent standards and methods for testing, evaluating, and certifying their performance, reliability, and security. For example, AVs can undergo various tests and simulations, in controlled or real-world environments, to assess and demonstrate their capabilities and limitations, and to identify and address any potential issues or failures.

- Technological trust can also be built or improved among the devices or systems involved in AVs, by providing and promoting clear and informative communication and feedback mechanisms for the users and the public, to increase their awareness and understanding of the technology and its implications. For example, AVs can display or explain their actions and intentions, such as their route, speed, or maneuvers, to the users and other road users, to reduce their uncertainty and anxiety, and to increase their confidence and acceptance.

In this section we analyzed the social and ethical implications of autonomous vehicles, and how they impact or improve the quality of life and well-being of the drivers, passengers, pedestrians, and other road users. We also discussed the effects of autonomous vehicles on the safety, efficiency, and accessibility of transportation systems and services, and the trade-offs and choices that they involve. We also investigated the environmental, energy, and emission impacts and challenges of autonomous vehicles, and how they challenge or contribute to the sustainability and resilience of the planet and the society. We also explored the behavior, preferences, and values of the human agents involved or affected by autonomous vehicles, and how they shape or respond to the design and development of autonomous vehicles. We also examined the ethical, legal, and moral dilemmas and questions that autonomous vehicles raise, such as responsibility, liability, and trust, and how they affect or ensure the accountability and transparency of autonomous vehicles. These are some of the social and ethical implications of autonomous vehicles and how they influence or enhance the quality of life and well-being of the users and the society. However, these implications are not simple or straightforward, but rather complex and nuanced, depending on various factors and perspectives, such as human factors, environmental factors, ethical principles, and moral values. Therefore, it is important to analyze and evaluate the social and ethical implications of autonomous vehicles, and to address and resolve the issues and challenges that they raise.

SECTION 4

The Policy and Regulatory Framework for Autonomous Vehicles

In this section, we will review the policy and regulatory framework for autonomous vehicles, and how they vary across different regions and jurisdictions, such as local, national, and international levels. We will also identify the main challenges and opportunities for policy-making and regulation of autonomous vehicles, and the goals and objectives that they aim to achieve. We will also survey some of the existing and proposed laws and standards for autonomous vehicles, and how they address the technical, economic, social, and ethical issues and challenges of autonomous vehicles, such as vehicle safety, road traffic, data protection, cyber security, and liability. We will also provide some of the best practices and recommendations for designing and implementing effective and inclusive policies and regulations for autonomous vehicles, and how they align with the vision and goals of the 2030 Agenda for Sustainable Development, adopt a human-centric and ethical approach to the development and deployment of autonomous vehicles, engage and consult with the relevant and diverse stakeholders and experts in the policy-making and regulation process, adapt and innovate the policies and regulations to the specific contexts and needs of different regions and jurisdictions, and monitor and evaluate the policies and regulations to the outcomes and impacts of autonomous vehicles.

What are the main challenges and opportunities for policy-making and regulation of autonomous vehicles at local, national, and international levels?

Autonomous vehicles (AVs) pose significant challenges and opportunities for policy-making and regulation, as they involve complex and interrelated technical, economic, social, and ethical issues, and affect multiple and diverse stakeholders, such as manufacturers, operators, users, regulators, and society. Policy-making and regulation of AVs require a coordinated and adaptive approach, that can balance the trade-offs and conflicts among different values and interests, such as safety, efficiency, fairness, and innovation, and that can address the uncertainties and risks associated with the development and deployment of AVs. Some of the main challenges and opportunities for policy-making and regulation of AVs at local, national, and international levels are:

Local Level: The local level refers to the sub-national or municipal level of policy-making and regulation, such as cities, counties, or states. The local level faces the following challenges and opportunities for AVs:

- **Challenges:** The local level faces the challenges of integrating and coordinating AVs with the existing and emerging transportation systems and services, such as public transportation, shared mobility, and micro-mobility, and of ensuring the accessibility and inclusivity of AVs for all, especially for the underserved and disadvantaged groups, such as the elderly, the disabled, the young, and the low-income. The local level also faces the challenges of managing and regulating the impacts and externalities of AVs on the local environment, infrastructure, land use, and public health, such as emissions, congestion, noise, parking, and safety.

- **Opportunities:** The local level has the opportunities to leverage and promote AVs as a means to achieve the local goals and visions for sustainable and smart urban development and mobility, such as reducing greenhouse gas emissions, improving air quality, enhancing road safety, increasing road capacity, optimizing traffic flow, and providing mobility and independence for all. The local level also has the opportunities of experimenting and innovating with AVs, by creating and testing various pilot projects, programs, and policies, that can foster the public acceptance and trust of AVs, and that can generate valuable data and insights for the evaluation and improvement of AVs.

National Level: The national level refers to the federal or central level of policy-making and regulation, such as countries or regions. The national level faces the following challenges and opportunities for AVs:

- **Challenges:** The national level faces the challenges of establishing and harmonizing the legal and regulatory frameworks and standards for AVs, that can provide clarity and consistency for the development and deployment of AVs across different jurisdictions and sectors, and that can address the legal and ethical issues and dilemmas raised by AVs, such as responsibility, liability, and trust. The national level also faces the challenges of supporting and stimulating the innovation and competitiveness of the AV industry and market, by providing incentives and funding for research and development, facilitating collaboration and cooperation among different stakeholders, and creating a level playing field and a conducive environment for AVs.

- **Opportunities:** The national level has the opportunity to lead and shape the vision and direction for the future of mobility and society with AVs, by setting the goals and priorities for the advancement and adoption of AVs, by engaging and consulting with the public and the experts, and by developing and implementing the policies and strategies for AVs. The national level also has the opportunity of benefiting and learning from the experiences and best practices of other countries or regions, by participating and collaborating in the international dialogues and initiatives on AVs, and by benchmarking and adapting the international standards and guidelines for AVs.

International Level: The international level refers to the global or transnational level of policy-making and regulation, such as organizations, agreements, or networks. The international level faces the following challenges and opportunities for AVs:

- **Challenges:** The international level faces the challenges of coordinating and aligning the policies and regulations of AVs among different countries or regions, that may have different legal systems, cultural values, and development stages, and that may have different interests and perspectives on AVs.

The international level also faces the challenges of addressing and resolving the cross-border and global issues and impacts of AVs, such as trade, security, privacy, and ethics, that may require common or compatible rules and standards, and that may involve conflicts or disputes among different parties.

- **Opportunities:** The international level has the opportunities to facilitate and advance the global development and deployment of AVs, by providing a platform and a forum for the exchange and dissemination of information, knowledge, and best practices on AVs, by fostering the cooperation and collaboration among different stakeholders, and by promoting the harmonization and convergence of the policies and regulations of AVs. The international level also has the opportunity to influence and contribute to the global goals and agendas for sustainable and inclusive development and mobility with AVs, such as the United Nations Sustainable Development Goals, the Paris Agreement, and the New Urban Agenda.

What are the existing and proposed laws and standards for autonomous vehicles and how do they vary across different regions and jurisdictions?

Autonomous vehicles (AVs) are subject to various laws and standards that regulate their development and deployment, such as vehicle safety, road traffic, data protection, cyber security, and liability. However, the existing and proposed laws and standards for AVs are not uniform or consistent across different regions and jurisdictions, as they reflect the different legal systems, cultural values, and development stages of the countries or regions involved. Some of the existing and proposed laws and standards for AVs and how they vary across different regions and jurisdictions are:

Vehicle Safety: Vehicle safety refers to the technical and functional requirements and specifications that ensure the safety and reliability of the vehicles and their components, such as sensors, computers, communication systems, and navigation systems. Vehicle safety can be regulated by national or international authorities, such as the National Highway Traffic Safety Administration (NHTSA) in the United States, the European Union Agency for the Cooperation of Energy Regulators (ACER) in the European Union, or the United Nations Economic Commission for Europe (UNECE) in the United Nations. Some of the existing and proposed laws and standards for vehicle safety and how they vary across different regions and jurisdictions are:

- **Existing Laws and Standards:** The existing laws and standards for vehicle safety are mainly based on the assumption that the vehicles are driven by human drivers and that the drivers are responsible for the control and supervision of the vehicles. Therefore, the existing laws and standards for vehicle safety focus on the physical and mechanical aspects of the vehicles, such as the brakes, steering, lights, and airbags, and the human-machine interface aspects of the vehicles, such as the dashboard, pedals, and seatbelts. The existing laws and standards for vehicle safety also vary in their scope and level of detail, depending on the region and jurisdiction involved. For example, the United States has a self-certification system, where the manufacturers are responsible for ensuring that their vehicles comply with the federal safety standards, while the European Union has a type-approval system, where the vehicles have to be approved by a designated authority before they can be sold or registered.

- **Proposed Laws and Standards:** The proposed laws and standards for vehicle safety are aimed at addressing the specific challenges and opportunities posed by AVs, such as the automation and connectivity of the vehicles, and the transfer and delegation of control and supervision from the human drivers to the vehicles. Therefore, the proposed laws and standards for vehicle safety focus on the software and system aspects of the vehicles, such as the perception, planning, and decision-making processes, and the communication and coordination with other vehicles and infrastructures. The proposed laws and standards for vehicle safety also aim to harmonize and converge the different regional and jurisdictional approaches, by adopting and adapting the international frameworks and guidelines, such as the UNECE's World Forum for Harmonization of Vehicle Regulations (WP.29), which has adopted several regulations and amendments for AVs, such as the Regulation on Automated Lane Keeping Systems (ALKS), the Regulation on Event Data Recorders (EDR), and the Regulation on Software Updates and Software Update Management Systems (SUMS).

Road Traffic: Road traffic refers to the rules and regulations that govern the operation and behavior of the vehicles and their drivers on the road, such as speed limits, traffic signals, lane markings, and road signs. Road traffic can be regulated by national or local authorities, such as the Department of Transportation (DOT) in the United States, the Ministry of Transport (MOT) in the United Kingdom, or the Road Traffic Authority (RTA) in

Australia. Some of the existing and proposed laws and standards for road traffic and how they vary across different regions and jurisdictions are:

- **Existing Laws and Standards:** The existing laws and standards for road traffic are mainly based on the assumption that the vehicles are driven by human drivers and that the drivers are responsible for the compliance and enforcement of the road traffic rules and regulations. Therefore, the existing laws and standards for road traffic focus on the licensing and registration of vehicles and their drivers, the testing and training of the drivers, and the penalties and sanctions for the violations of the road traffic rules and regulations. The existing laws and standards for road traffic also vary in their scope and level of detail, depending on the region and jurisdiction involved. For example, the

United States has a federal-state system, where the federal government sets the minimum standards for the licensing and registration of vehicles and their drivers, while the states have the authority to set their own specific rules and regulations for the road traffic, such as the speed limits, the traffic signals, and the lane markings.

- **Proposed Laws and Standards:** The proposed laws and standards for road traffic are aimed at addressing the specific challenges and opportunities posed by AVs, such as the interaction and coordination of the vehicles with other road users, such as pedestrians, cyclists, and other vehicles, and the adaptation and modification of the road traffic rules and regulations to accommodate the capabilities and limitations of the AVs. Therefore, the proposed laws and standards for road traffic focus on the testing and certification of the AVs and their systems, the operation and supervision of the AVs on the road, and the liability and insurance of the AVs and their operators. The proposed laws and standards for road traffic also aim to harmonize and converge the different regional and jurisdictional approaches, by adopting and adapting the international frameworks and guidelines, such as the Vienna Convention on Road Traffic, which has been amended to allow the use of AVs, as long as they can be overridden or switched off by the driver, and the Geneva Convention on Road Traffic, which has been amended to allow the use of AVs, as long as they comply with the road traffic rules and regulations.

Data Protection: Data protection refers to the rules and regulations that protect the privacy and security of the personal and sensitive data that are collected, processed, and transmitted by the vehicles and their systems, such as the location, speed, and destination of the vehicles, and the identity, preferences, and behavior of the drivers and passengers. Data protection can be regulated by national or international authorities, such as the Federal Trade Commission (FTC) in the United States, the General Data Protection Regulation (GDPR) in the European Union, or the Organisation for Economic Co-operation and Development (OECD) in the Organisation for Economic Co-operation and Development. Some of the existing and proposed laws and standards for data protection and how they vary across different regions and jurisdictions are:

- **Existing Laws and Standards:** The existing laws and standards for data protection are mainly based on the assumption that the data are collected, processed, and transmitted by human agents, and that the data subjects have the rights and control over their data, such as the right to access, rectify, erase, or object to their data. Therefore, the existing laws and standards for data protection focus on the principles and obligations for the data controllers and processors, such as the consent, purpose, and transparency of the data collection, processing, and transmission, and the security, accountability, and compliance of the data protection measures. The existing laws and standards for data protection also vary in their scope and level of detail, depending on the region and jurisdiction involved. For example, the United States has a sectoral and fragmented system, where the data protection laws and standards are different for different sectors and states, such as the Health Insurance Portability and Accountability Act (HIPAA) for the health sector, the Children's Online Privacy Protection Act (COPPA) for the children's sector, and the California Consumer Privacy Act (CCPA) for the state of California, while the European Union has a comprehensive and harmonized system, where the data protection laws and standards are the same for all sectors and member states, such as the General Data Protection Regulation (GDPR), which applies to all personal data that are collected, processed, or transmitted within or outside the European Union.

- **Proposed Laws and Standards:** The proposed laws and standards for data protection are aimed at addressing the specific challenges and opportunities posed by AVs, such as the volume and variety of the data that are collected, processed, and transmitted by the vehicles and their systems, and the value and vulnerability of the data for the users and the society. Therefore, the proposed laws and standards for data protection focus on the rights and responsibilities of the data subjects and stakeholders, such as the drivers, passengers, manufacturers, operators, regulators, and third parties, and the principles and guidelines for data governance and management, such as the minimization, anonymization, and aggregation of the data, and the sharing, access, and use of the data. The proposed laws and standards for data protection also aim to harmonize and converge the different regional and jurisdictional approaches, by adopting and adapting the international frameworks and guidelines, such as the OECD Guidelines on the Protection of Privacy and Transborder

Flows of Personal Data, which provide the basic principles and recommendations for the data protection and the cross-border data flows, and the OECD Principles on Artificial Intelligence, which provide the human-centric and ethical principles and recommendations for the data protection and the AI development and deployment.

Cyber Security: Cyber security refers to the rules and regulations that protect the integrity and availability of the systems and networks that enable or support the operation or function of AVs, such as the sensors, computers, communication systems, or navigation systems, and that prevent or mitigate the cyber threats and attacks that may compromise or damage the systems and networks, such as hacking, spoofing, or jamming. Cybersecurity can be regulated by national or international authorities, such as the Department of Homeland Security

(DHS) in the United States, the European Union Agency for Cybersecurity (ENISA) in the European Union, or the International Organization for Standardization (ISO) in the International Organization for Standardization. Some of the existing and proposed laws and standards for cyber security and how they vary across different regions and jurisdictions are:

- **Existing Laws and Standards:** The existing laws and standards for cyber security are mainly based on the assumption that the systems and networks are operated and maintained by human agents and that the cyber threats and attacks are detected and responded to by human agents. Therefore, the existing laws and standards for cyber security focus on the principles and obligations of the system and network operators and providers, such as the identification, assessment, and mitigation of cyber risks, and the notification, reporting, and recovery of cyber incidents. The existing laws and standards for cyber security also vary in their scope and level of detail, depending on the region and jurisdiction involved. For example, the United States has a voluntary and collaborative system, where the system and network operators and providers are encouraged to share information and best practices on cyber security with the federal government and other stakeholders, such as the Automotive Information Sharing and Analysis Center (Auto-ISAC), while the European Union has a mandatory and prescriptive system, where the system and network operators and providers are required to comply with the minimum standards and obligations on cyber security, such as the

Directive on Security of Network and Information Systems (NIS Directive).

- **Proposed Laws and Standards:** The proposed laws and standards for cyber security are aimed at addressing the specific challenges and opportunities posed by AVs, such as the complexity and interdependency of the systems and networks, and the autonomy and connectivity of the vehicles. Therefore, the proposed laws and standards for cyber security focus on the rights and responsibilities of the system and network users and stakeholders, such as the drivers, passengers, manufacturers, operators, regulators, and third parties, and the principles and guidelines for the system and network design and development, such as the security by design, security by default, and security by update. The proposed laws and standards for cyber security also aim to harmonize and converge the different regional and jurisdictional approaches, by adopting and adapting the international frameworks and guidelines, such as the ISO/SAE 21434 standard on Road Vehicles Cybersecurity Engineering, which provides the requirements and recommendations for the cybersecurity lifecycle of AVs, and the UNECE's WP.29 regulation on Cyber Security and Cyber Security Management System, which provides the requirements and procedures for the type approval and market surveillance of AVs.

Liability: Liability refers to the legal obligation or duty to pay for or remedy the damages or losses caused by one's actions, especially when they are negligent or wrongful. Liability can be civil, criminal, or administrative, depending on the nature and severity of the damages or losses, and the type and level of the legal authority involved. AVs can raise questions and issues about the determination and allocation of liability among different agents, such as drivers, passengers, manufacturers, operators, regulators, and third parties, and the criteria and mechanisms for doing so. Some of the existing and proposed laws and standards for liability and how they vary across different regions and jurisdictions are:

- **Existing Laws and Standards:** The existing laws and standards for liability are mainly based on the assumption that the vehicles are driven by human drivers and that the drivers are liable for the damages or losses caused by their actions, especially when they are negligent or wrongful. Therefore, the existing laws and standards for liability focus on the principles and rules for the attribution and imputation of liability among different agents, such as the fault, causation, and foreseeability of the actions and outcomes, and the compensation and punishment of liability among different agents, such as the damages, remedies, and penalties for the actions and outcomes. The existing laws and standards for liability also vary in their scope and level of detail, depending on the region and jurisdiction involved. For example, the United States has a tort-based system, where the liability is determined by the civil courts based on the common law or the statutory law, while the European Union has a no-fault system, where the liability is determined by the insurance companies based on the compulsory motor insurance or the product liability directive.

- **Proposed Laws and Standards:** The proposed laws and standards for liability are aimed at addressing the specific challenges and opportunities posed by AVs, such as the shift and share of liability from the human drivers to the vehicles and their systems, and the complexity and uncertainty of the actions and outcomes of AVs. Therefore, the proposed laws and standards for liability focus on the principles and rules for the allocation and transfer of liability among different agents, such as the level, mode, and situation of the automation and the control of the vehicles, and the compensation and settlement of liability among different agents, such as the insurance, warranty, and contract for the vehicles and their systems. The proposed laws and standards for liability also aim to harmonize and converge the different regional and jurisdictional approaches, by adopting and adapting the international frameworks and guidelines, such as the UNECE's WP.29 regulation on Automated Driving Systems (ADS), which provides the requirements and procedures for the type approval and market surveillance of AVs, and the European Commission's Expert Group on Liability and New Technologies, which provides the recommendations and guidelines for the liability and new technologies, such as AVs.

What are the best practices and recommendations for designing and implementing effective and inclusive policies and regulations for autonomous vehicles?

Based on the web search results, some of the best practices and recommendations for designing and implementing effective and inclusive policies and regulations for autonomous vehicles (AVs) are:

- Aligning the policies and regulations with the vision and goals of the 2030 Agenda for Sustainable Development, which emphasizes the importance of inclusive development and the reduction of inequalities. AVs can contribute to the achievement of several Sustainable Development Goals (SDGs), such as SDG 3 (Good Health and Well-being), SDG 7 (Affordable and Clean

Energy), SDG 9 (Industry, Innovation and Infrastructure), SDG 11 (Sustainable Cities and Communities), and SDG 13 (Climate Action), by improving the safety, efficiency, and accessibility of transportation systems and services, and by reducing the environmental and social impacts and externalities of transportation.

- Adopting a human-centric and ethical approach to the development and deployment of AVs, which respects and protects the rights, dignity, and diversity of all human agents involved or affected by AVs, such as drivers, passengers, pedestrians, and other road users. AVs should be designed and operated in a way that ensures the safety, security, privacy, and well-being of the human agents, and that prevents or mitigates any potential harm or discrimination that may arise from the use of AVs.

- Engaging and consulting with the relevant and diverse stakeholders and experts in the policy-making and regulation process, such as manufacturers, operators, users, regulators, researchers, civil society, and the public. AVs involve complex and interrelated technical, economic, social, and ethical issues and challenges, that require the participation and collaboration of different actors and perspectives, to ensure the legitimacy, effectiveness, and sustainability of the policies and regulations.

- Adapting and innovating the policies and regulations to the specific contexts and needs of different regions and jurisdictions, such as cities, countries, or regions. AVs have different impacts and implications for different settings and situations, that depend on various factors, such as the level of development, the infrastructure, the culture, and the legal system. Therefore, the policies and regulations should be tailored and customized to the local conditions and realities, while also learning and benefiting from the experiences and best practices of other regions and jurisdictions.

- Monitoring and evaluating the policies and regulations to the outcomes and impacts of AVs, such as the safety, efficiency, and accessibility of transportation systems and services, and the environment, energy consumption, and emissions. AVs are dynamic and evolving technologies that pose new opportunities and challenges for policy-making and regulation. Therefore, the policies and regulations should be based on evidence and data, and should be reviewed and revised regularly, to ensure their relevance, effectiveness, and accountability.

In this section we reviewed the policy and regulatory framework for autonomous vehicles, and how they vary across different regions and jurisdictions, such as local, national, and international levels. We also identified the main challenges and opportunities for policy-making and regulation of autonomous vehicles, and the goals and objectives that they aim to achieve. We also surveyed some of the existing and proposed laws and standards for autonomous vehicles, and how they address the technical, economic, social, and ethical issues and challenges of autonomous vehicles, such as vehicle safety, road traffic, data protection, cyber security, and liability. We also provided some of the best practices and recommendations for designing and implementing effective and inclusive policies and regulations for autonomous vehicles, and how they align with the vision and goals of the 2030 Agenda for Sustainable Development, adopt a human-centric and ethical approach to the development and deployment of autonomous vehicles, engage and consult with the relevant and diverse stakeholders and experts in the policy-making and regulation process, adapt and innovate the policies and regulations to the specific contexts and needs of different regions and jurisdictions, and monitor and evaluate the policies and regulations to the outcomes and impacts of autonomous vehicles. These are some of the policy and regulatory aspects of autonomous vehicles and how they affect or ensure the safety, efficiency, accessibility, and sustainability of the transportation system and service. However, these aspects are not uniform or consistent, but rather diverse and dynamic, depending on various factors and conditions, such as technology, market, policy, and society. Therefore, it is important to design and implement effective and inclusive policies and regulations for autonomous vehicles, and to align and coordinate them with the vision and goals of the 2030 Agenda for Sustainable Development, the human-centric and ethical approach to the development and deployment of autonomous vehicles, and the relevant and diverse stakeholders and experts in the policy-making and regulation process.

Conclusion

In this book, we have explored the various aspects and dimensions of autonomous vehicles (AVs), such as the history, technology, applications, impacts, and challenges of AVs. We have also discussed the social and ethical implications of AVs, such as the safety, efficiency, and accessibility of transportation systems and services, the environment, energy consumption, and emissions, the behavior, preferences, and values of drivers, passengers, and pedestrians, and the responsibility, liability, and trust among various stakeholders, such as manufacturers, operators, users, and regulators. Furthermore, we have examined the policy and regulatory framework for AVs, and how they vary across different regions and jurisdictions, such as local, national, and international levels. We have also reviewed some of the existing and proposed laws and standards for AVs, and how they address the technical, economic, social, and ethical issues and challenges of AVs, such as vehicle safety, road traffic, data protection, cyber security, and liability.

By the end of this book, you should have a comprehensive and critical understanding of the opportunities and challenges that AVs present for the future of mobility and society, and the need for a multidisciplinary and collaborative approach to address them. Some of the key implications and recommendations for various stakeholders, such as consumers, manufacturers, policymakers, and researchers, are:

Consumers: Consumers are the potential or actual users of AVs, such as drivers, passengers, pedestrians, and other road users. Consumers should be aware and informed of the benefits and risks of AVs, and the rights and responsibilities that they have when using or interacting with AVs. Consumers should also be involved and engaged in the development and deployment of AVs, by providing feedback and input on their needs, preferences, and values, and by participating in the testing and evaluation of AVs.

Manufacturers: Manufacturers are the producers and providers of AVs and their components, such as sensors, computers, communication systems, and navigation systems. Manufacturers should be innovative and competitive in the design and development of AVs, and ensure that their products and services meet the highest standards of quality, safety, reliability, and security. Manufacturers should also be accountable and transparent in the operation and maintenance of AVs, and provide clear and accurate information and documentation on their products and services, and the data and software that they collect, process, and transmit.

Policymakers: Policymakers are the authorities and institutions that govern, regulate, or provide the systems and services related to AVs, such as public transportation, shared mobility, and micro-mobility. Policymakers should be proactive and adaptive in the establishment and harmonization of the legal and regulatory frameworks and standards for AVs, and address the legal and ethical issues and dilemmas raised by AVs, such as responsibility, liability, and trust. Policymakers should also be supportive and stimulating in the promotion and adoption of AVs, provide incentives and funding for research and development, and facilitate collaboration and cooperation among different stakeholders.

Researchers: Researchers are the experts and professionals who conduct and disseminate the scientific and technical knowledge and understanding of AVs, such as engineers, computer scientists, psychologists, and sociologists. Researchers should be rigorous and creative in the exploration and investigation of the various aspects and dimensions of AVs, and generate valuable data and insights for the evaluation and improvement of AVs. Researchers should also be ethical and responsible in the conduct and communication of their research, and respect and protect the rights, dignity, and diversity of the human agents involved or affected by AVs.

However, despite the progress and achievements that have been made in the field of AVs, there are still many limitations and gaps in the current knowledge and understanding of AVs, and many directions for future research and development. Some of the main limitations and gaps are:

Technical Limitations and Gaps: The technical limitations and gaps refer to the challenges and difficulties that AVs face in achieving and maintaining the high level of performance, reliability, and security that are required for their safe and efficient operation and function, such as the perception, planning, and decision-making processes, and the communication and coordination with other vehicles and infrastructures. The technical limitations and gaps also refer to the lack and inconsistency of the data and methods that are used for the testing, evaluation, and certification of AVs, such as the scenarios, metrics, and criteria that are used to assess and demonstrate the capabilities and limitations of AVs.

Economic Limitations and Gaps: The economic limitations and gaps refer to the barriers and uncertainties that AVs face in achieving and maintaining the high level of innovation and competitiveness that are required for their successful and sustainable development and deployment, such as the cost, availability, and accessibility of the technology and the infrastructure that is needed for AVs, and the demand, acceptance, and adoption of the products and services that are offered by AVs. The economic limitations and gaps also refer to the lack and inconsistency of the models and methods that are used for the analysis, prediction, and optimization of the impacts and outcomes of AVs, such as the costs, benefits, and externalities that are associated with AVs.

Social Limitations and Gaps: The social limitations and gaps refer to the conflicts and trade-offs that AVs face in achieving and maintaining the high level of alignment and harmony that are required for their integration and coordination with the existing and emerging transportation systems and services, such as the public transportation, shared mobility, and micro-mobility, and the behavior, preferences, and values of the drivers, passengers, pedestrians, and other road users. The social limitations and gaps also refer to the lack and inconsistency of the theories and methods that are used for the understanding, explanation, and intervention of the social and ethical implications of AVs, such as the safety, efficiency, and accessibility of transportation systems and services, and the responsibility, liability, and trust among various stakeholders.

Policy Limitations and Gaps: The policy limitations and gaps refer to the delays and discrepancies that AVs face in achieving and maintaining the high level of clarity and consistency that are required for their compliance and enforcement of the legal and regulatory frameworks and standards for AVs, such as the vehicle safety, road traffic, data protection, cyber security, and liability. The policy limitations and gaps also refer to the lack and inconsistency of the frameworks and guidelines that are used for the design and implementation of the policies and regulations for AVs, such as the vision, goals, and priorities for the development and deployment of AVs, and the engagement and consultation of the stakeholders and experts in the policy-making and regulation process.

Some of the directions for future research and development are:

Technical Directions: The technical directions refer to the improvement and enhancement of the performance, reliability, and security of AVs, such as the development and integration of advanced sensors, computers, communication systems, and navigation systems, and the implementation and optimization of intelligent and adaptive perception, planning, and decision-making algorithms, and the communication and coordination protocols with other vehicles and infrastructures.

Economic Directions: The economic directions refer to the assessment and evaluation of the innovation and competitiveness of AVs, such as the estimation and comparison of the costs, benefits, and externalities of AVs, and the identification and analysis of the factors and drivers that influence the demand, acceptance, and adoption of AVs, and the development and application of the models and methods that can predict and optimize the impacts and outcomes of AVs.

Social Directions: The social directions refer to the exploration and investigation of the alignment and harmony of AVs, such as the understanding and explanation of the behavior, preferences, and values of the drivers, passengers, pedestrians, and other road users, and the intervention and influence of the social and ethical implications of AVs, such as the safety, efficiency, and accessibility of transportation systems and services, and the responsibility, liability, and trust among various stakeholders.

Policy Directions: The policy directions refer to the establishment and harmonization of the legal and regulatory frameworks and standards for AVs, such as the development and adaptation of the laws and standards that address the legal and ethical issues and dilemmas raised by AVs, such as vehicle safety, road traffic, data protection, cyber security, and liability, and the design and implementation of the policies and regulations that promote and support the development and deployment of AVs, such as the vision, goals, and priorities for the future of mobility and society with AVs, and the engagement and consultation of the stakeholders and experts in the policy-making and regulation process.

We hope that this book has provided you with a useful and insightful overview of the various aspects and dimensions of AVs, and has stimulated your interest and curiosity for further learning and research on this fascinating and important topic. We also hope that this book has inspired you to think critically and creatively about the opportunities and challenges that AVs present for the future of mobility and society and to contribute to the multidisciplinary and collaborative approach that is needed to address them.

Acknowledgment

I would like to express my sincere gratitude and appreciation to all the people who have contributed to the creation and completion of this book. Without their support and guidance, this book would not have been possible.

First and foremost, I would also like to thank my family, friends, and colleagues, for their constant love, support, and inspiration. They have been my motivation and my strength, and they have shared with me their insights and experiences in options trading.

I would like to thank Bing, who assisted me in the creation of my book. Bing was a reliable, helpful, and friendly source of information and inspiration.

Last but not least, I would like to thank you, the reader, for choosing this book, and for joining me on exploring automated vehicles. I hope that this book will provide you with a solid foundation and a valuable guide on the ever evolving technology.

I hope you found my service helpful and valuable. I would love to hear your honest feedback and ratings, as they will help me improve my service quality and customer satisfaction. Please let me know if you have any suggestions or comments on how I can serve you better.

Roman Preciado